max maths primary

A SINGAPORE APPROACH

Journal 3

ALICE HANSEN

AF891379

macmillan education
INTERNATIONAL CURRICULUM

Macmillan Education Limited
4 Crinan Street
London, N1 9XW

Companies and representatives throughout the world

www.macmillanic.com

ISBN 978-1-380-02482-4

Text © Macmillan Education Limited 2018
Design and illustration © Macmillan Education Limited and Star Publishing Pte Ltd
Written by Alice Hansen

The author has asserted her right to be identified as the author of this work in accordance with the Copyright, Design and Patents Act 1988.

First published 2018

All rights reserved. No part of this publication may be reproduced, stored in a retrieval system, or transmitted in any form, or by any means, electronic, mechanical, photocopying, recording, or otherwise, without the prior written permission of the publishers.

Original design by John Dickinson and Macmillan Education Limited
Page layout by Tek-Art
Illustrations by Robbie Short and Mark Ruffle c/o Beehive Illustration Agency
Picture research by Fiona Gartland/Out of House
Cover design by Macmillan Education Limited
Cover illustration by Star Publishing Pte Ltd
Cover photography by Shutterstock/3d_kot
Cover picture research by Susannah Jayes

Printed and bound in Spain

2028 2027 2026 2025 2024

22 21 20 19 18 17 16

Acknowledgements

The authors and publishers would like to thank the following for permission to reproduce their images:

p12 PhotoDisc (orange butterfly), PhotoDisc/Getty Images (brown striped butterfly), PhotoDisc (brown butterfly), Macmillan Mexico/Tony Sanchez (turquoise butterfly), Getty (green butterfly), Getty (blue butterfly), Getty (yellow butterfly); **p13** ImageSource; **p46** Getty Images/iStockphoto / Thinkstock/ Davidenko Andrey (beehive), PhotoDisc/Getty Images (door), Getty (moon); **p58** PhotoDisc (butterfly), PhotoDisc/Getty Images (ladybird), PhotoDisc (leaf), Getty Images/eli_asenova (orange), Getty Images/iStockphoto/Knaupe (pinecone); **p93** Getty Images/iStockphoto/Thinkstock/ene (bird), Brand X (gift box); **p96** PhotoDisc (window and door), Getty Images/Thinkstock/Hemera Technologies (set square); **p98** Getty Images/Thinkstock/Hemera Technologies

Contents

Helping your child at home

Welcome to Max Maths Journal 3! This journal is designed to help you support your child with their mathematical learning, including the language of mathematics. It can be used either as part of the Max Maths scheme or as a standalone resource.

This journal provides opportunities for consolidation of school learning and reflection, so it is important to make sure your child has covered a topic at school before your child begins work on it in the journal. Typically, your child will be set exercises from the journal as homework by their teacher.

We recommend that when your child sits down to work on the journal, you are on hand to provide support, engage in discussion and explore the maths together. Some tasks in the journal require the direct involvement of a grown-up for discussion and you are expected to reflect with your child on their understanding at the end of each topic.

Topic structure

Each topic begins with a scenario and related task that reflects key learning from the topic.

Check your maths!

These tasks check your child's understanding of one or more of the key concepts in the topic.

Practising my maths language

These activities are focused on practising new mathematical language connected to the topic. They often require discussion with an adult at home.

From school to home

Opportunities for discussion, sharing and reflection on your child's learning. At the end of the topic, there is space for your child's teacher to comment on your child's learning experience.

1 Numbers up to 1 000

Counting to 1 000

We have been counting up to 200 and beyond.

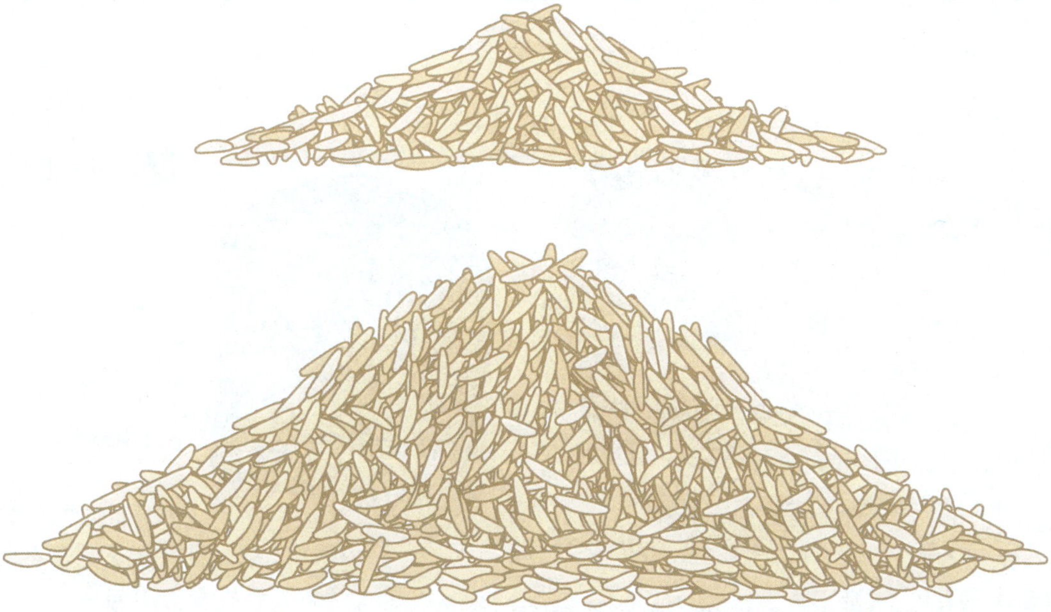

Ask to use a bag of rice from your kitchen.

Make a pile of exactly 200 grains of rice by counting the grains.

Make another pile of rice that has approximately 1 000 grains in it. Do not count them all. Use what you know about 200 and 1 000 to estimate what the pile would look like.

Check your maths!

The children are counting in ones from 167, but only one of them has counted correctly.

a Which child is correct?

b What errors have the other children made?

Practising my
maths language

Look at these numbers.

a Choose a number from the group below and count on 20 more.

b Which will give you the biggest number when you have counted on 20 more? Show an adult and tell them why.

c Which number will give you the smallest? Show an adult and tell them why.

d Does the adult agree with your answers to **b** and **c**?

125 592

98 801

203 177

From school to home

Find someone at home to talk to.

1. **Show them your school work. Talk about some of your answers to the questions on numbers up to 1 000.**
2. **Explain what you have learnt about counting to 1 000.**
3. **Ask someone at home to help you complete the following:**

a What do you like about counting to 1 000?

____

____

____

b What was hard about counting to 1 000?

____

____

____

I understand

I understand a bit

I need more time to understand this

4 **How well do you understand this topic now? Circle one face for each statement.**

- I can count to 200 and beyond.
- I can read numbers to 1 000.
- I can write numbers to 1 000.

5 **Ask an adult at home to read and sign this.**

I have checked learning on this topic.

We have shared some understanding at home.

Signed ______________________________

Date ____________

Teacher comment ______________________

____

Date ____________

Counting within 1 000

We have been using our place-value knowledge to count on and back in steps of 1, 10 and 100. Toby and Han are playing a counting game that helps them practise their counting.

Play the same game as Toby and Han with someone at home. Take turns to choose any number and then count on or count back. Try counting on and back in steps of 1, then 10 and then 100. Each time the other person should try to work out what your pattern is.

Check your maths!

1 What numbers come next?

a 338, 438, 538, 638, ☐, ☐, ☐

b 562, 552, ☐, 532, 522, ☐, ☐, ☐

c 54, 53, 52, 51, ☐, ☐, ☐

d 276, 286, ☐, ☐, 316, ☐, 336

2 Make up your own number patterns that count on or count back in steps of 1, 10 or 100 and write them below. You could draw a number line or place-value counters to help you.

Practising my

maths language

Complete the sentences using the words below.

one ten one hundred more less

a 583 is ten more than 573.

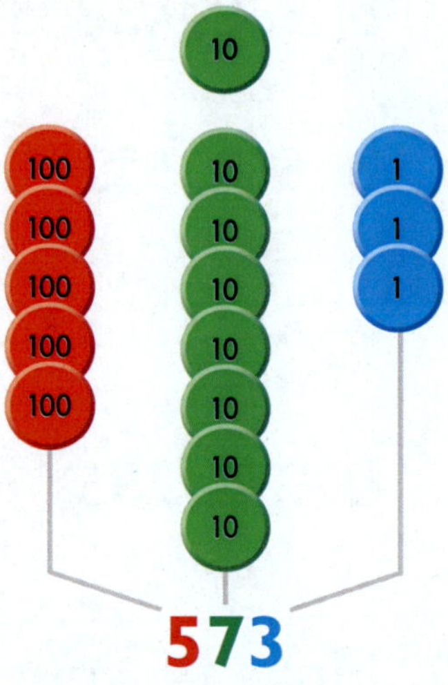

b 279 is ________ than 280.

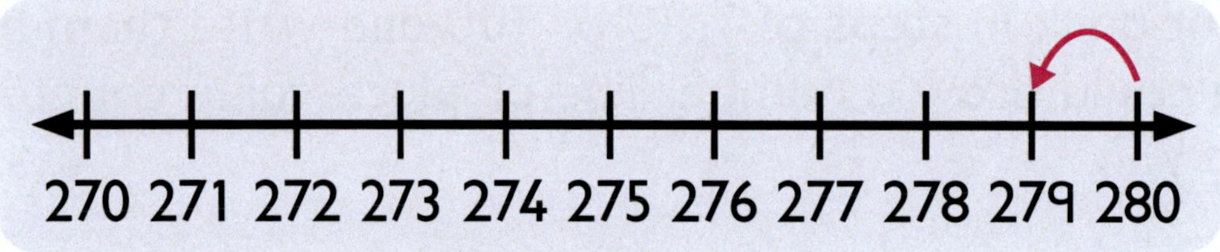

c 590 is ________ than 690.

d 192 is ________ than 92.

e 916 is ________ than 906.

f 628 is ________ than 638.

g 1 000 is ________ than 999.

h 821 is ________ than 921.

From school to home

Find someone at home to talk to.

1. **Show them your school work. Talk about some of your answers to the questions on numbers up to 1 000.**
2. **Explain what you have learnt about counting within 1 000.**
3. **Ask someone at home to help you complete the following:**

a What do you like about counting within 1 000?

___

___

___

b What was hard about counting within 1 000?

___

___

___

I understand | I understand a bit | I need more time to understand this

4 How well do you understand this topic now? Circle one face for each statement.

- I can complete a number pattern that counts on or counts back in steps of 1, 10 and 100.
- I can make up my own number patterns that count on or count back in steps of 1, 10 and 100.
- I can find 1, 10 or 100 more or less than another number.

5 Ask an adult at home to read and sign this.

I have checked learning on this topic.

We have shared some understanding at home.

Signed ________________________________

Date ____________

Teacher comment ________________________

____

Date ____________

Estimating

We have been grouping objects in tens to make a sensible estimate of a number as a range.

Do you agree with Samir's or Padma's estimate?

Tell an adult and explain how you worked out your estimate. Now count how many buttons there are. Are you surprised by who was correct?

Check your maths!

Estimate how many butterflies are in this group.

Estimate range

Actual number

Practising my maths language

Work with someone at home to find some small objects (such as grains of rice, buttons, pebbles or pieces of pasta).

a Estimate how many objects there are.

I estimate there are ______.

b How many does the person you are working with estimate there are?

c Count the objects. There are ______.

d Who made the closer estimate? ______

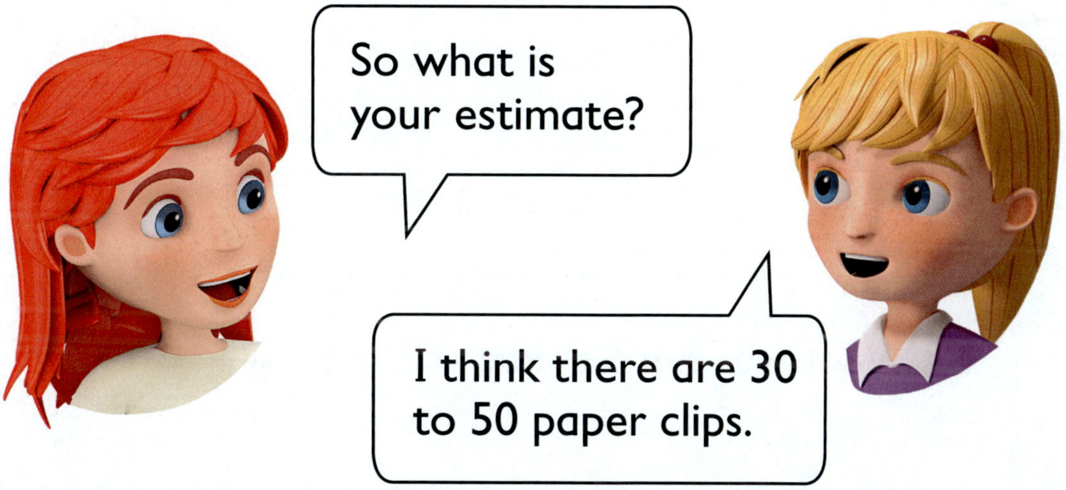

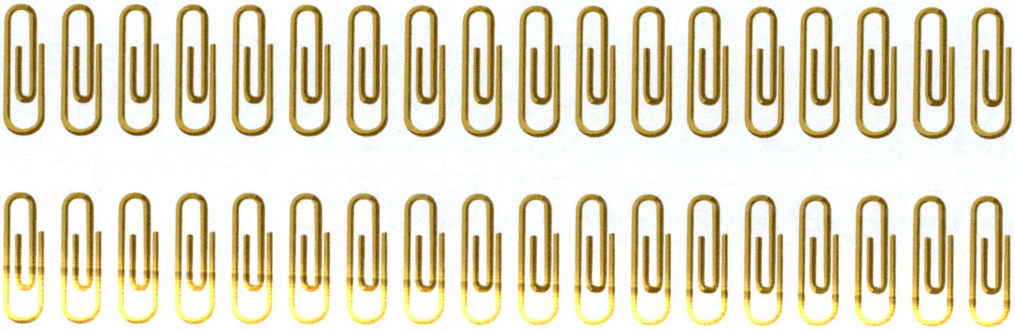

From school to home

Find someone at home to talk to.

1. **Show them your school work. Talk about some of your answers to the questions on numbers up to 1 000.**
2. **Explain what you have learnt about estimating.**
3. **Ask someone at home to help you complete the following:**

a What do you like about estimating?

___

___

___

b What was hard about estimating?

___

___

___

I understand

I understand a bit

I need more time to understand this

4 **How well do you understand this topic now? Circle one face for each statement.**

- I can make sensible estimates of a number of objects as a range.
- I can count a number of objects to check my estimates.

5 **Ask an adult at home to read and sign this.**

I have checked learning on this topic.

We have shared some understanding at home.

Signed ________________________________

Date ____________

Teacher comment ________________________

____

Date ____________

2 Addition

Mental addition

We have been using our place-value knowledge to add 10s and 100s to numbers mentally.

Look at the abacus in the last picture and use it to explain to someone at home how the children would add 300 to 292.

Check your maths!

Complete the following addition sentences.

a 274 + 20 = ☐

b 381 + 400 = ☐

c 36 + ☐ = 86

d 925 + ☐ = 955

e ☐ + 500 = 728

f ☐ + 30 = 271

Remember to imagine an abacus or place-value counters if it helps you to calculate the sum.

Practising my maths language

Look at these addition sentences. There is an error in each one.

Explain to an adult at home why the sums are incorrect.

Remember to use your place-value knowledge, including **ones**, **tens** and **hundreds**. Use the space below for your workings.

a 149 + 30 = 449

b 134 + 50 = 634

c 57 + 200 = 77

d 427 + 100 = 428

From school to home

Find someone at home to talk to.

1. **Show them your school work. Talk about some of your answers to the questions on addition.**
2. **Explain what you have learnt about addition using 10s and 100s.**
3. **Ask someone at home to help you complete the following:**

a What do you like about addition using 10s and 100s?

____

____

____

b What was hard about addition using 10s and 100s?

____

____

____

I understand

I understand a bit

I need more time to understand this

4 How well do you understand this topic now? Circle one face for each statement.

- I can add 10 and multiples of 10 to 2- and 3-digit numbers.
- I can add 100 and multiples of 100 to 3-digit numbers.
- I can use my place-value knowledge to explain how to add 10s and 100s mentally.

5 Ask an adult at home to read and sign this.

I have checked learning on this topic.

We have shared some understanding at home.

Signed ______________________________

Date ____________

Teacher comment ______________________

____

Date ____________

Adding 2-digit numbers

We have been learning how to add pairs of 2-digit numbers mentally by regrouping the numbers into tens and ones.

Hmmm, how can I work out the sum?

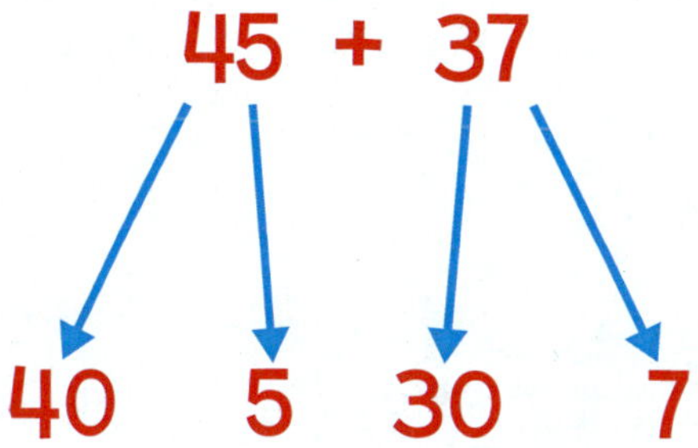

$$45 + 37 = 40 + 30 + 5 + 7$$
$$= 70 + 12$$
$$= 82$$

So, 45 + 37 = 82

Explain to an adult what Padma has done to add together 45 and 37. Use the words **tens**, **ones** and **regroup**.

Check your maths!

Find the sums of these pairs of numbers by first regrouping the numbers into tens and ones.

a 65 + 22 = ☐

b 78 + 41 = ☐

c 42 + 98 = ☐

d 73 + 44 = ☐

e 85 + 67 = ☐

f 23 + 63 = ☐

Practising my maths language

Ask someone at home to give you two 2-digit numbers to add together mentally. You can make notes to show how you are regrouping the numbers in the space below. Talk about what you are doing. Remember to use terms such as **tens**, **ones** and **regroup** to help your explanations. Then try another one.

Here is what you might say for 53 + 25.

"I have to add 53 + 25.

So, in my head I regroup the numbers into tens and ones.

Now I add 50 and 20, which is equal to 70.

Next I add 3 and 5, which is equal to 8.

Finally, I add 70 and 8, which is equal to 78.

So, 53 + 25 = 78."

53 + 25

50 3 20 5

$$\begin{aligned} 53 + 25 &= 50 + 20 + 3 + 5 \\ &= 70 + 8 \\ &= 78 \end{aligned}$$

So, 53 + 25 = 78.

From school to home

Find someone at home to talk to.

1. **Show them your school work. Talk about some of your answers to the questions on addition.**
2. **Explain what you have learnt about adding 2-digit numbers.**
3. **Ask someone at home to help you complete the following:**

a What do you like about adding 2-digit numbers?

____

____

____

b What was hard about adding 2-digit numbers?

____

____

____

I understand | I understand a bit | I need more time to understand this

4 How well do you understand this topic now? Circle one face for each statement.

- I can regroup numbers into tens and ones.
- I can add pairs of 2-digit numbers by regrouping mentally.

5 Ask an adult at home to read and sign this.

I have checked learning on this topic.

We have shared some understanding at home.

Signed ______________________________

Date __________

Teacher comment ______________________

____

Date __________

3 Subtraction

Mental addition and subtraction

Han and Samir are adding or subtracting multiples of 10 from 2- and 3-digit numbers.

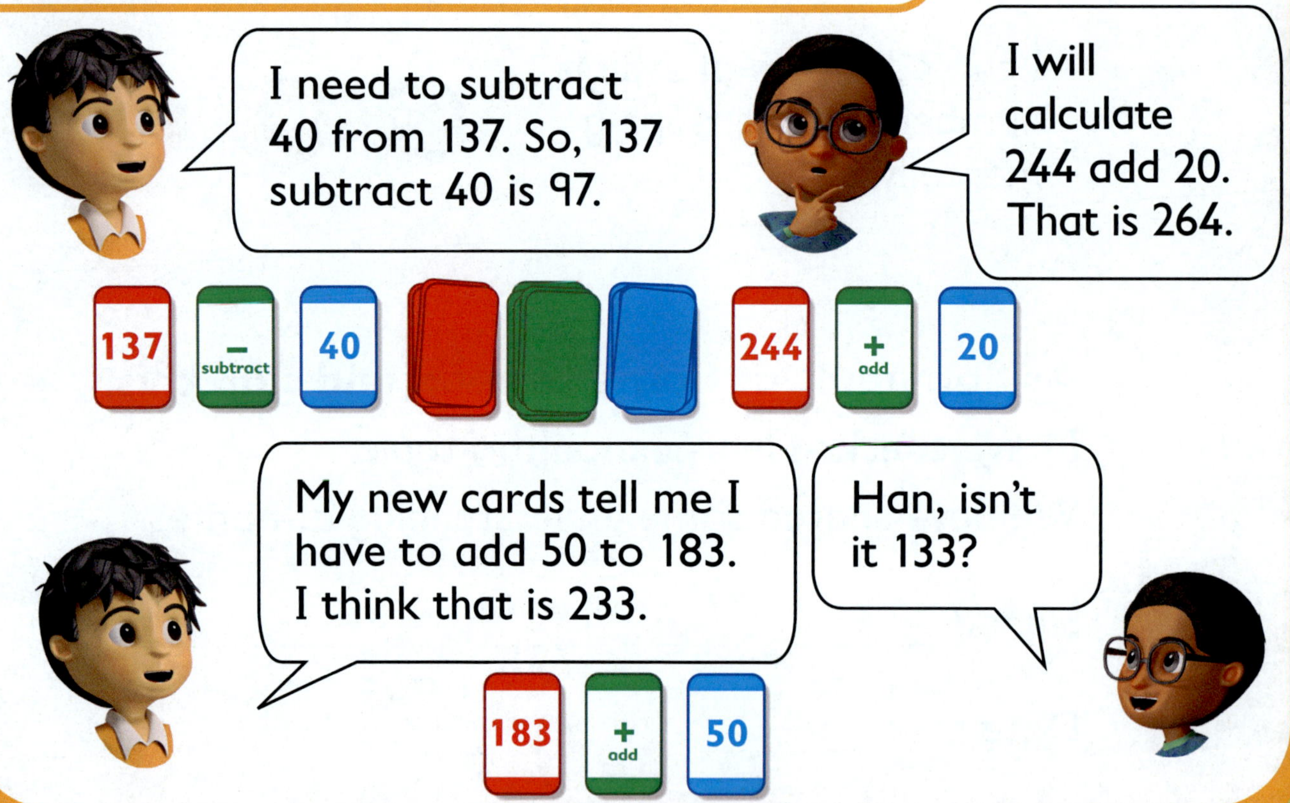

Who is correct? What do you think the child who has the wrong solution did? Can you demonstrate the correct answer using a number line?

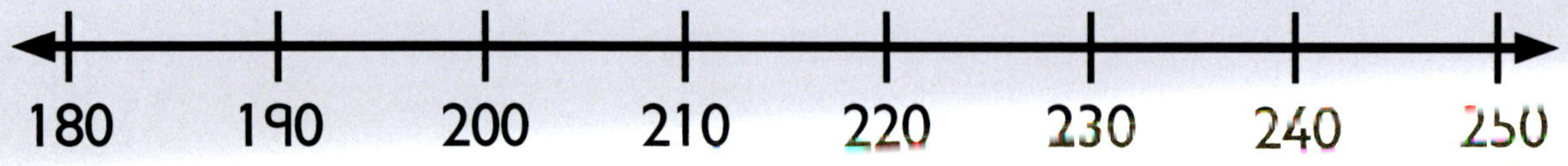

Check your maths!

Imagine you are playing the game on the previous page.
What would be your solutions for these cards?

a 104 − subtract 60

b 52 − subtract 30

c 290 + add 70

d 683 + add 10

e 902 − subtract 30

f 251 + add 40

Practising my maths language

Explain what is happening on each number line using the correct language. Choose from **count on**, **count back**, **add**, **subtract** to use in your sentences.

a

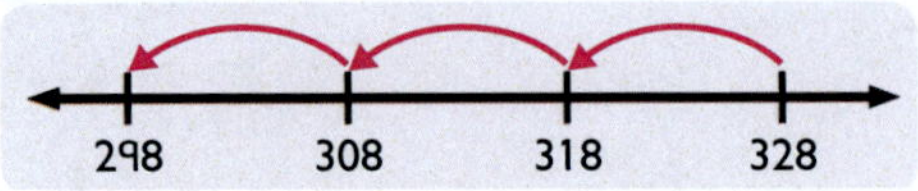

I can count back 3 tens to subtract 30 from 328.
328 - 30 = 298

b

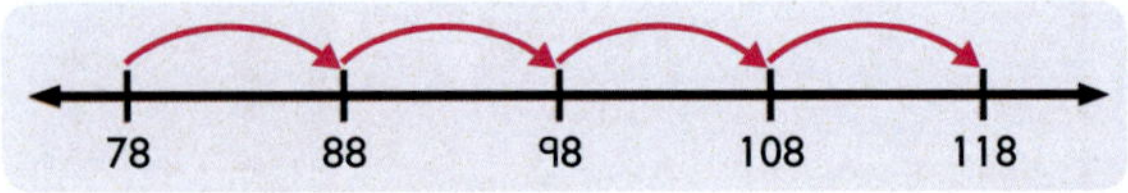

c

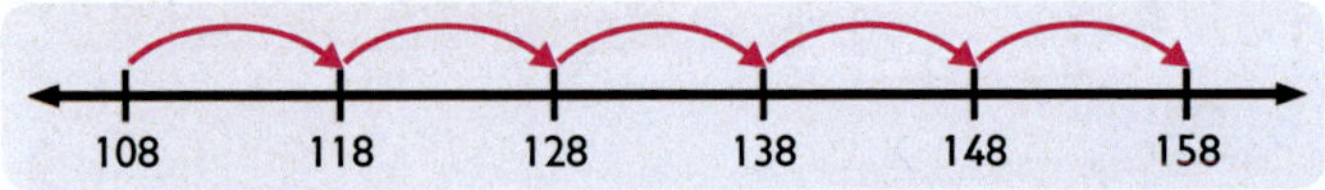

d

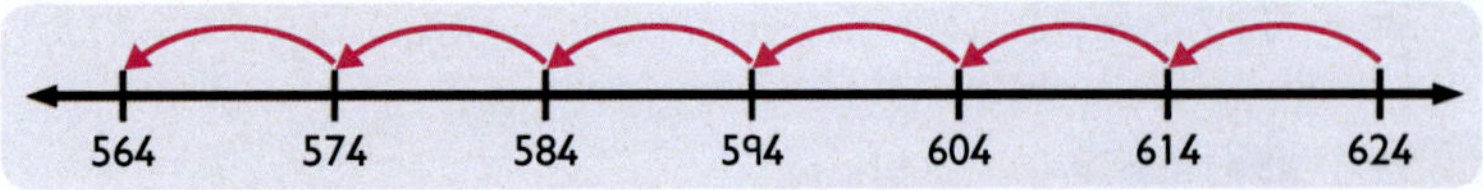

e

From school to home

Find someone at home to talk to.

1. **Show them your school work. Talk about some of your answers to the questions on subtraction.**
2. **Explain what you have learnt about mental subtraction.**
3. **Ask someone at home to help you complete the following:**

a What do you like about mental subtraction?

____

____

____

b What was hard about mental subtraction?

____

____

____

I understand

I understand a bit

I need more time to understand this

4 How well do you understand this topic now? Circle one face for each statement.

- I can count on in multiples of 10.
- I can count back in multiples of 10.
- I can use counting on and back to add and subtract multiples of 10 from 2- and 3-digit numbers.

5 Ask an adult at home to read and sign this.

I have checked learning on this topic.

We have shared some understanding at home.

Signed ______________________________

Date __________

Teacher comment ______________________

____

Date __________

Subtracting 2-digit numbers

We have used counting blocks to help us subtract numbers, by crossing off the number of blocks we are subtracting.

The teacher has asked the learners to write down what these counting blocks represent.

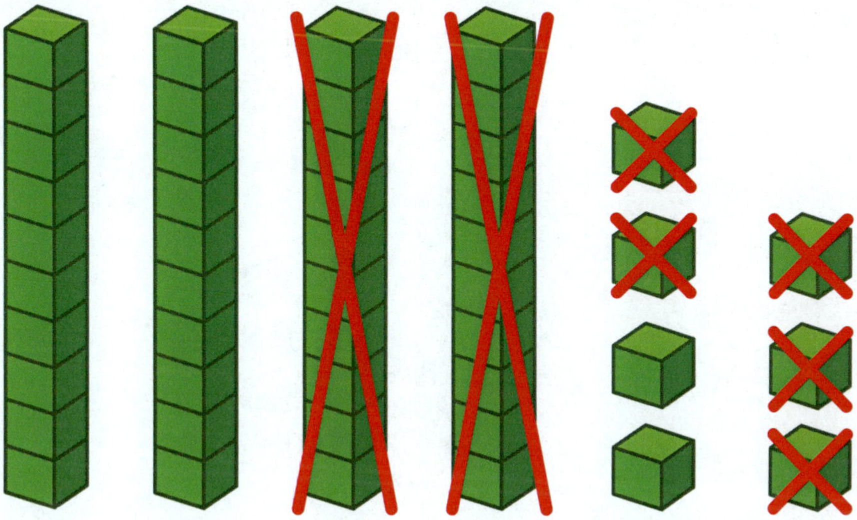

There are several ideas from the learners.

22 – 25 47 – 22 47 – 25 25 – 22

Which response do you think is correct? Why?

____

____

____

Check your maths!

Cross out the correct number of counting blocks to find the solution for these subtraction questions.

a 58 – 35

58 – 35 = ☐

b 27 – 24

27 – 24 = ☐

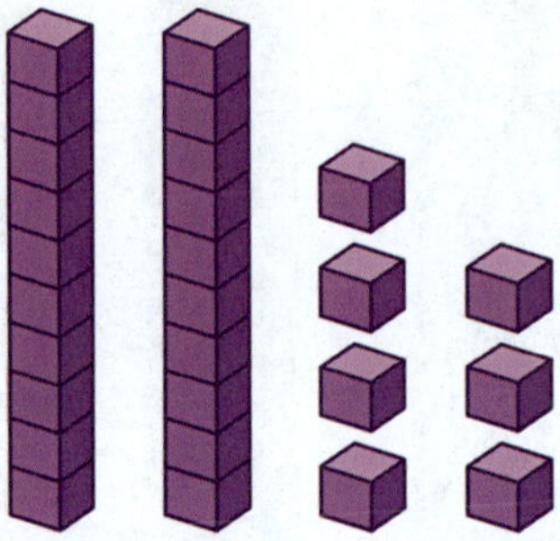

c 36 – 12

36 – 12 = ☐

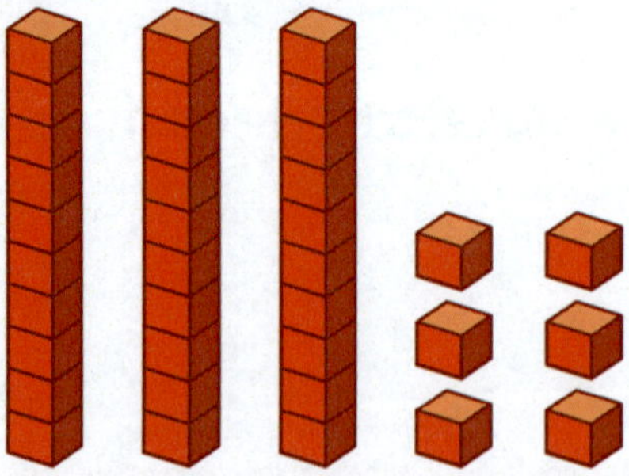

d 62 – 42

62 – 42 = ☐

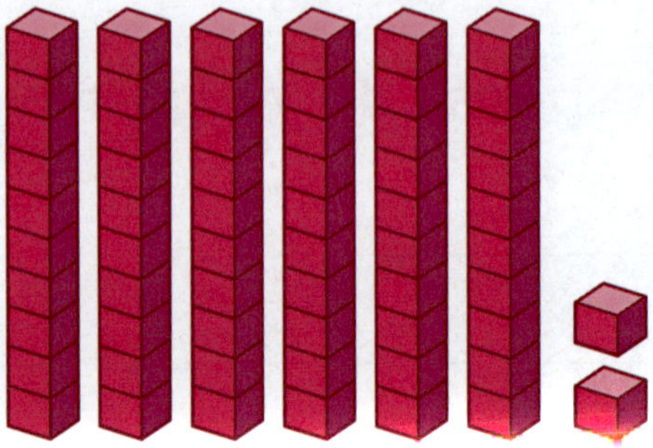

Practising my

maths language

Here are some words you will hear in maths lessons on addition and subtraction.

Circle each of the words that mean the same as the word **subtract**. Clue: there are five of them!

take away

less

hide

remove

add

more

altogether

plus

minus

From school to home

Find someone at home to talk to.

1. **Show them your school work. Talk about some of your answers to the questions on subtraction.**
2. **Explain what you have learnt about using counting blocks to subtract 2-digit numbers.**
3. **Ask someone at home to help you complete the following:**

a What do you like about using counting blocks to subtract 2-digit numbers?

____

____

____

b What was hard about using counting blocks to subtract 2-digit numbers?

____

____

____

I understand | I understand a bit | I need more time to understand this

4 **How well do you understand this topic now? Circle one face for each statement.**

- I can subtract 2-digit numbers from another number.
- I can use counting blocks to show the answer to a subtraction question.

5 **Ask an adult at home to read and sign this.**

I have checked learning on this topic.

We have shared some understanding at home.

Signed ______________________

Date __________

Teacher comment ______________________

____

Date __________

4 Fractions

Mixed numbers and improper fractions

The children are finding the equivalent mixed numbers and improper fractions by shading shapes.

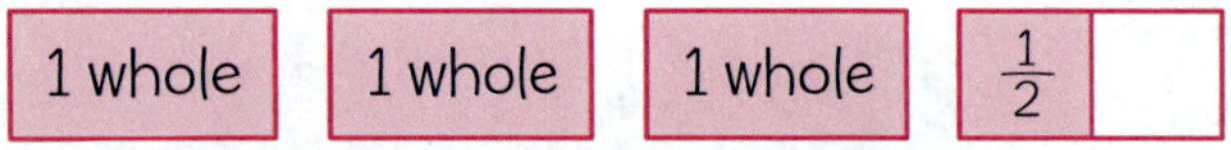

Toby says: "I have $3\frac{1}{2}$ blocks. I wonder how many halves this is equivalent to."

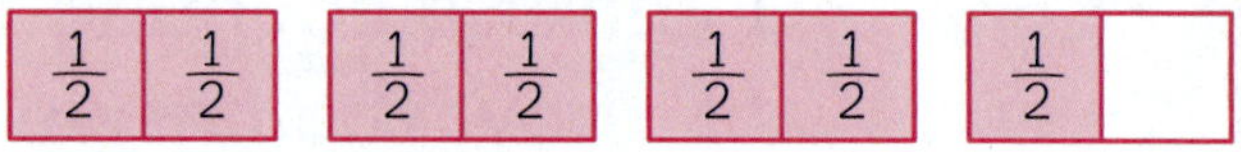

Tya says: "There are 7 halves in $3\frac{1}{2}$."

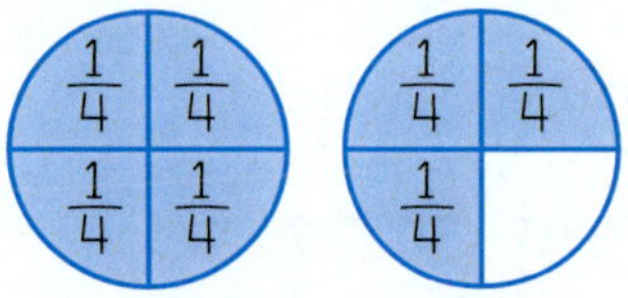

Jade says: "I have seven quarters. What is this equivalent to?"

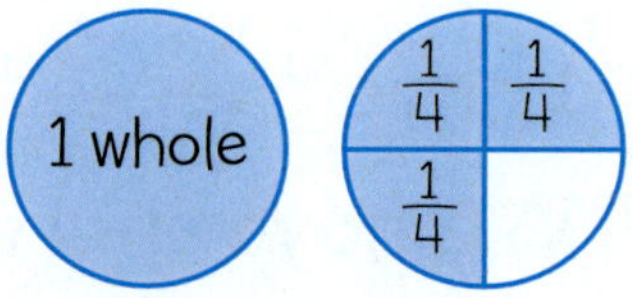

Samir says: "I think seven quarters is equivalent to one whole and three quarters."

Can you draw some shapes to show mixed numbers and equivalent improper fractions to show they are equivalent?

First, draw some shapes on a separate piece of paper. Then decide what fractions you are going to show. You could choose $\frac{1}{4}$s, $\frac{1}{2}$s, $\frac{1}{3}$s or even $\frac{1}{10}$s. Draw at least two sets of shapes using different fractions.

Check your maths!

1 Order these mixed numbers and simple fractions on the number line.

$\frac{1}{4}$ $\frac{1}{2}$ $1\frac{1}{2}$ $2\frac{1}{10}$ $3\frac{3}{4}$ $4\frac{1}{4}$

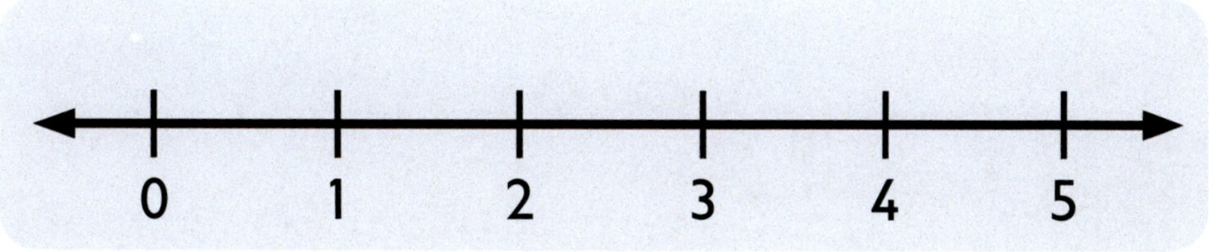

2 Order these mixed numbers and improper fractions on the number line.

$\frac{7}{4}$ $\frac{3}{2}$ $1\frac{1}{4}$ $3\frac{4}{10}$ $2\frac{3}{4}$ $2\frac{1}{3}$

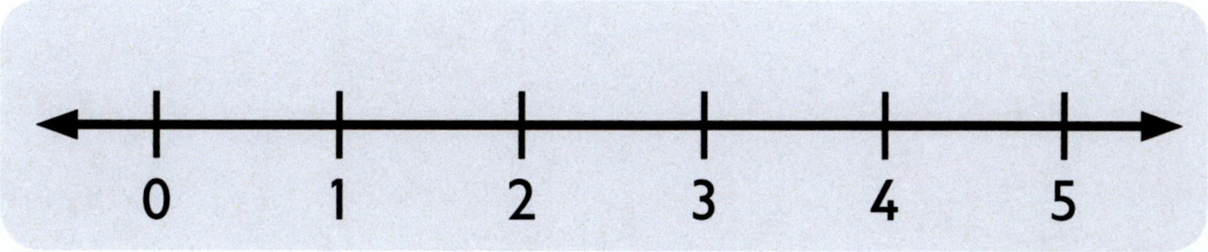

Practising my

maths language

Draw a circle around each **mixed number**, a square around each **proper fraction** and a triangle around each **improper fraction**.

$\frac{1}{2}$ $\frac{1}{10}$

$\frac{1}{4}$ $3\frac{1}{2}$

$\frac{10}{2}$ $\frac{1}{3}$

$\frac{3}{5}$

$6\frac{1}{4}$

$5\frac{1}{3}$ $\frac{4}{10}$

$\frac{8}{3}$

$2\frac{3}{4}$ $3\frac{1}{2}$

$\frac{5}{4}$

$\frac{12}{5}$ $8\frac{1}{10}$ $\frac{6}{4}$

From school to home

Find someone at home to talk to.

1. **Show them your school work. Talk about some of your answers to the questions on fractions.**
2. **Explain what you have learnt about mixed numbers and improper fractions.**
3. **Ask someone at home to help you complete the following:**

a What do you like about mixed numbers and improper fractions?

____

____

____

b What was hard about mixed numbers and improper fractions?

____

____

____

I understand

I understand a bit

I need more time to understand this

4 How well do you understand this topic now? Circle one face for each statement.

- I can identify proper fractions, improper fractions and mixed numbers.
- I can find equivalent mixed numbers and improper fractions by drawing shapes.
- I can order proper fractions and mixed numbers on a number line.

5 Ask an adult at home to read and sign this.

I have checked learning on this topic.

We have shared some understanding at home.

Signed ______________________________

Date ____________

Teacher comment ______________________

____

Date ____________

Equivalent fractions

Samir is exploring some equivalent fractions.

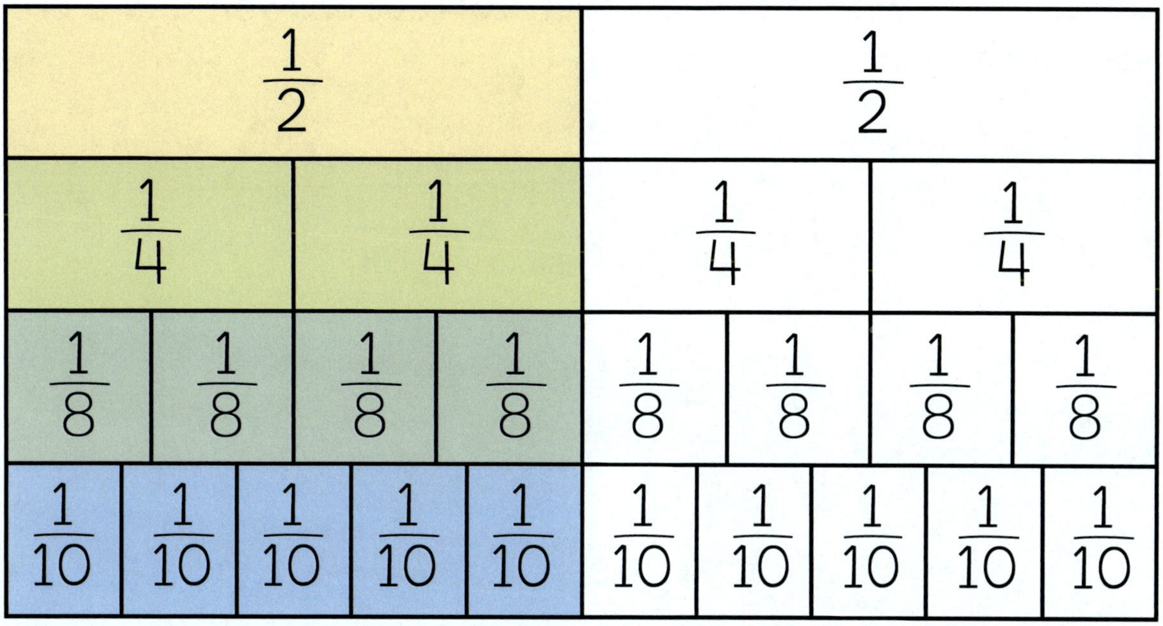

Samir says, "I can see that $\frac{1}{2}$, $\frac{2}{4}$, $\frac{4}{8}$ and $\frac{5}{10}$ are all equivalent fractions."

Draw a diagram like Samir's to show some fractions are equivalent.

Check your maths!

It's your turn to be the teacher! Mark this child's maths homework – with a tick for each correct answer, and a cross next to each incorrect answer.

$\frac{1}{2} = \frac{4}{8}$ ☐

$\frac{5}{10} = \frac{2}{4}$ ☐

$\frac{1}{2} = \frac{6}{10}$ ☐

$\frac{5}{10} = \frac{4}{8}$ ☐

$\frac{2}{4} = \frac{1}{2}$ ☐

$\frac{2}{4} = \frac{5}{8}$ ☐

$\frac{5}{10} = \frac{1}{2}$ ☐

$\frac{1}{2} = \frac{3}{4}$ ☐

$\frac{6}{10} = \frac{1}{2}$ ☐

$\frac{3}{7} = \frac{2}{4}$ ☐

Practising my maths language

Draw lines to match the fraction symbols to the correct words. There are more words than symbols, so some words will be left over.

	two fours
$\frac{1}{2}$	one half
	four eighths
$\frac{2}{4}$	fifteen
	one twoth
$\frac{5}{10}$	four eights
	two quarters
$\frac{4}{8}$	five tenths

From school to home

Find someone at home to talk to.

1. **Show them your school work. Talk about some of your answers to the questions on fractions.**
2. **Explain what you have learnt about equivalent fractions.**
3. **Ask someone at home to help you complete the following:**

a What do you like about equivalent fractions?

___

___

___

b What was hard about equivalent fractions?

___

___

___

I understand | I understand a bit | I need more time to understand this

4 **How well do you understand this topic now? Circle one face for each statement.**

- I can use representations to show fractions that are equivalent to one half.
- I can identify equivalent fractions.
- I can correctly read fraction symbols.

5 **Ask an adult at home to read and sign this.**

I have checked learning on this topic.

We have shared some understanding at home.

Signed ______________________________

Date __________

Teacher comment ______________________

____

Date __________

5 Shape and space

2D shapes

Han has been finding 2D shapes around his town.

These are regular hexagons.

This is an irregular pentagon.

This is a regular octagon.

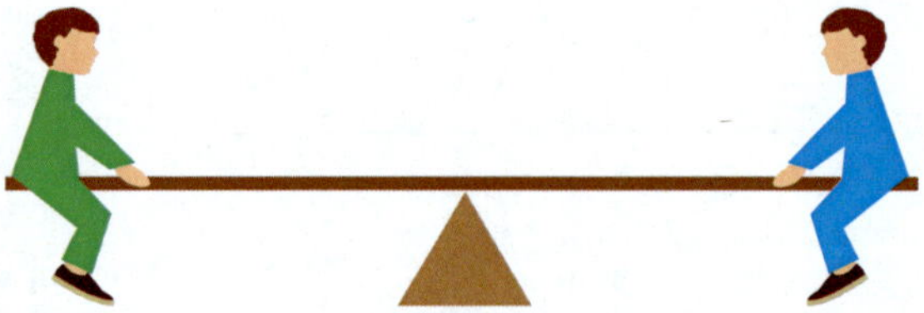

This is a triangle.

This is a rectangle.

This is a semicircle.

Can you find ten regular and irregular shapes in your home?

Check your maths!

Draw the following 2D shapes.

irregular hexagon

triangle

square

irregular octagon

regular pentagon

regular hexagon

irregular pentagon

regular octagon

rectangle

Practising my maths language

Write the name of each shape. Then write the properties of that shape. Use the terms below. The first one has been done for you.

right angle(s) **straight side(s)** **curved side(s)** **length**

2D shape	Properties
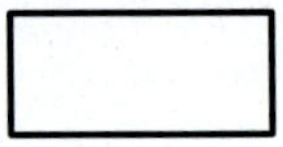	This is a rectangle. It has 4 vertices. It has 4 straight sides and 4 right angles.
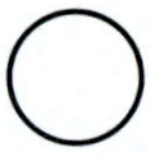	This is a ______. It has 0 vertices. It has no ______.
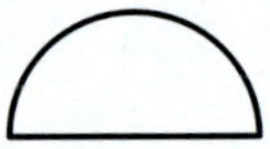	This is a ______. It has 2 vertices. It has 1 ______ and 1 ______.
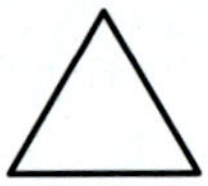	This is a ______. It has 3 vertices. It has 3 ______. The sides are of equal ______.
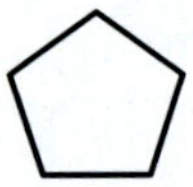	This is a ______. It has 5 vertices. It has 5 ______. The sides are of equal ______.
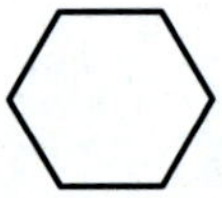	This is a ______. It has 6 vertices. It has 6 ______. The sides are of equal ______.
	This is an ______. It has 5 vertices. It has 5 ______. The sides are not of equal ______.
	This is a ______. It has 4 vertices. It has 4 ______ and 4 ______.
	This is a ______. It has 8 vertices. It has 8 ______. The sides are of equal ______.
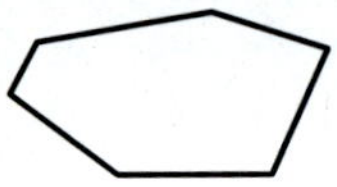	This is an ______. It has 6 vertices. The sides are not all of equal ______.

From school to home

Find someone at home to talk to.

1. **Show them your school work. Talk about some of your answers to the questions on shapes.**
2. **Explain what you have learnt about 2D shapes.**
3. **Ask someone at home to help you complete the following:**

a What do you like about 2D shapes?

____

____

____

b What was hard about 2D shapes?

____

____

____

I understand | I understand a bit | I need more time to understand this

4 How well do you understand this topic now? Circle one face for each statement.

- I can identify regular and irregular 2D shapes.
- I can draw regular and irregular 2D shapes.
- I can describe regular and irregular 2D shapes.

5 Ask an adult at home to read and sign this.

I have checked learning on this topic.

We have shared some understanding at home.

Signed ______________________________

Date __________

Teacher comment ______________________

____

Date __________

3D shapes

Jade, Tya and Padma are playing a game of 'Guess the shape'.

a

b

c

What are the three shapes in the bag?

a ______ **b** ______ **c** ______

Check your maths!

Look at these groups of objects. Why have they been grouped in this way?

a

They are all ______________________________

b

____

c

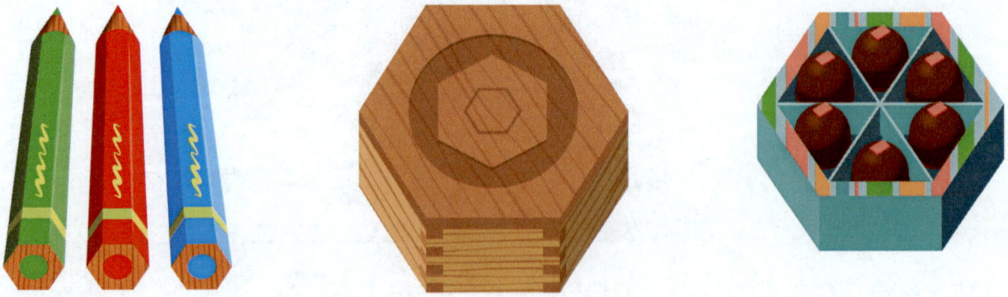

____

Practising my
maths language

1 Label the edge, face and vertex of each of these 3D shapes.

a

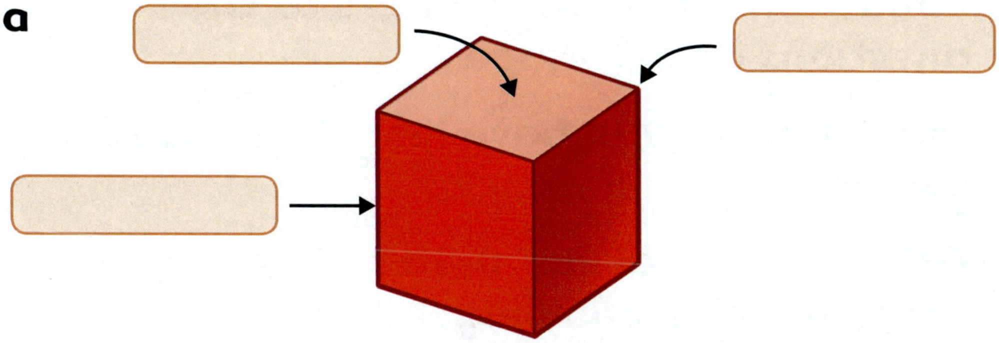

b

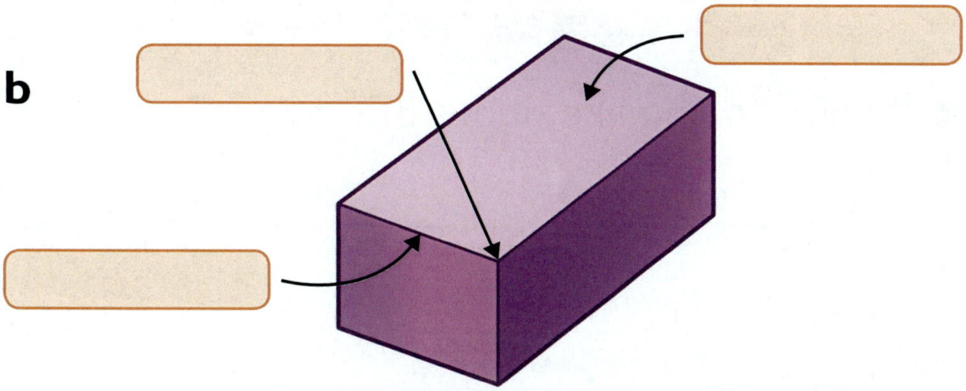

c

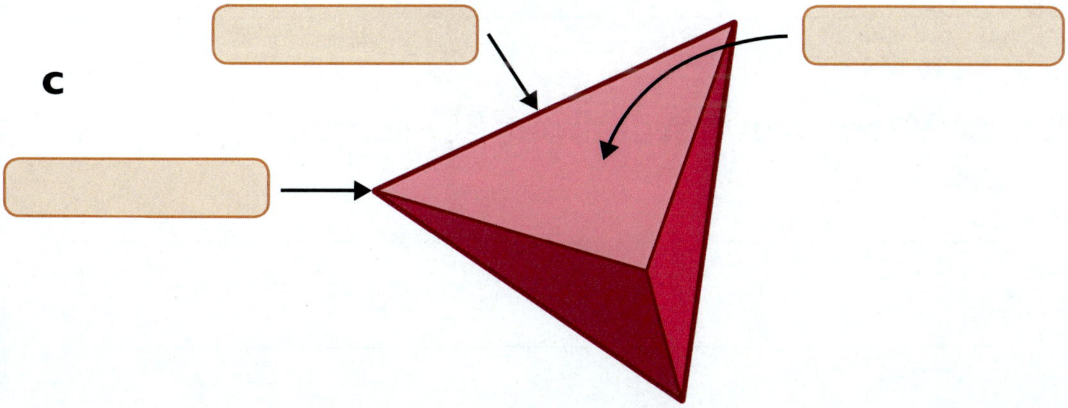

2 Look around you for some of the shapes on the Check your maths! page. Walk around your home with an adult and tell them the names of the shapes that you find.

From school to home

Find someone at home to talk to.

1. **Show them your school work. Talk about some of your answers to the questions on 3D shapes.**
2. **Explain what you have learnt about 3D shapes.**
3. **Ask someone at home to help you complete the following:**

a What do you like about 3D shapes?

____

____

____

b What was hard about 3D shapes?

____

____

____

I understand | I understand a bit | I need more time to understand this

4 **How well do you understand this topic now? Circle one face for each statement.**

- I can identify 3D shapes from everyday life.
- I can identify 3D shapes by using the number and shape of faces, number of vertices and edges.

5 **Ask an adult at home to read and sign this.**

I have checked learning on this topic.

We have shared some understanding at home.

Signed ______________________________

Date __________

Teacher comment ______________________

____

Date __________

Reflective symmetry

The children have been identifying lines of symmetry. Can you check their work to see if they have put the line of reflection in the right place? Are they all correct? Put a tick in the box next to the correct ones and a cross in the box by the incorrect ones.

a ☐

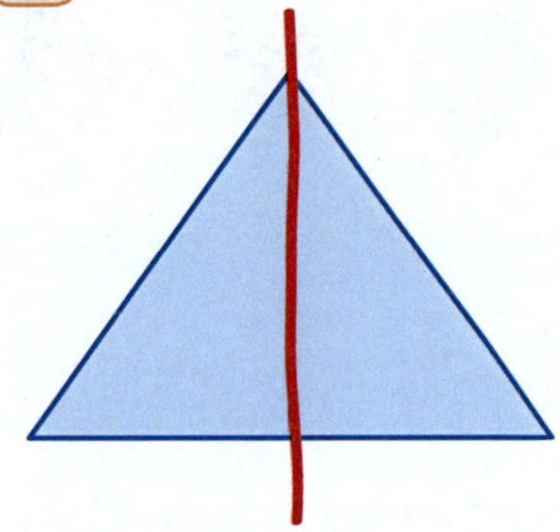

b ☐

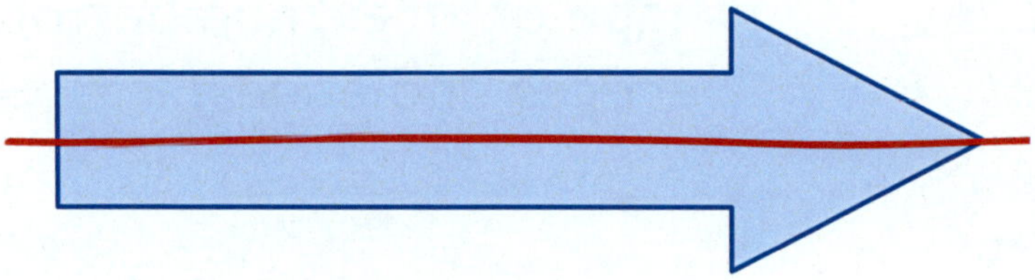

c ☐

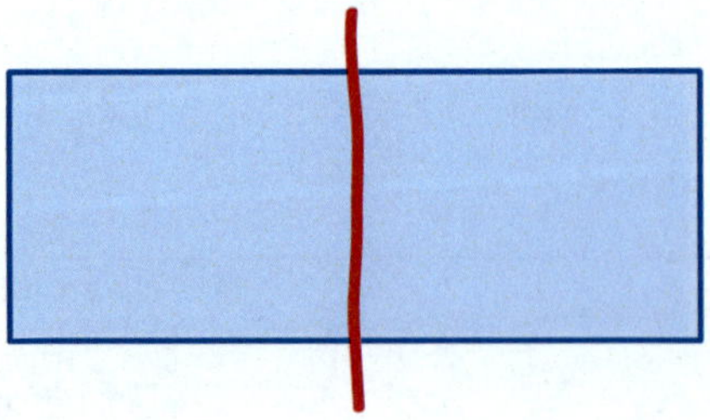

d ☐

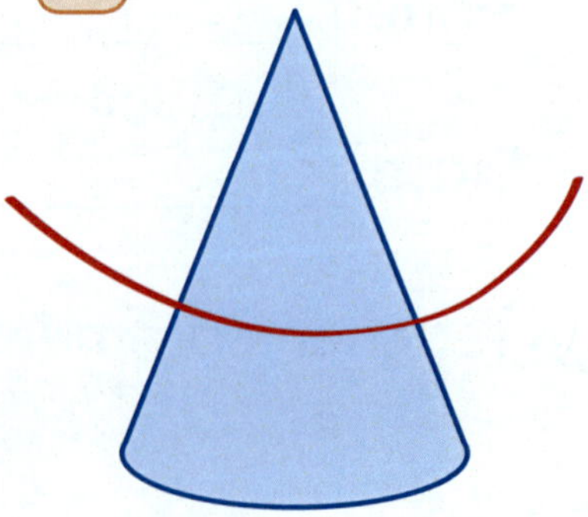

Check your maths!

Finish these shapes using the line of symmetry.

a

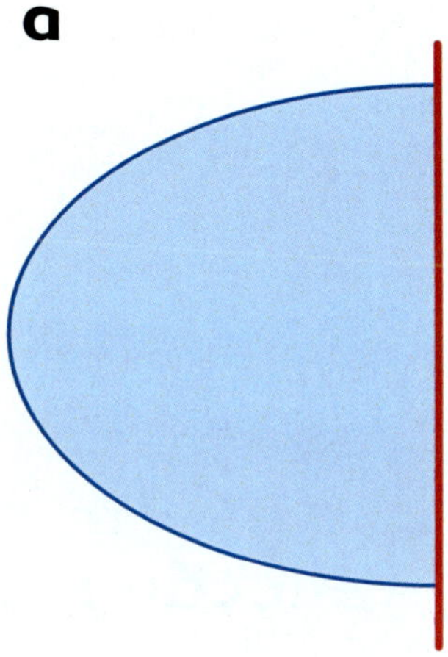

b

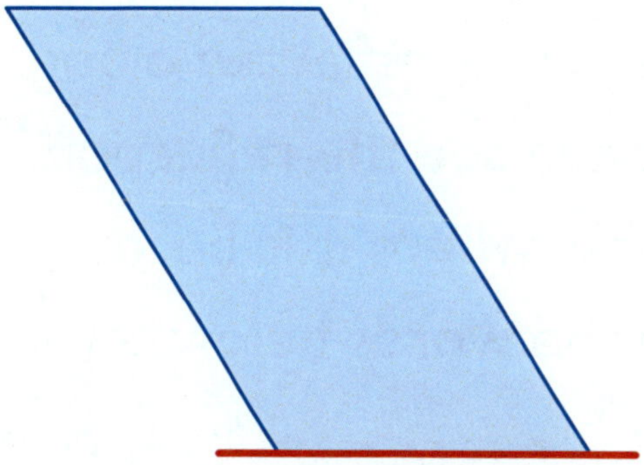

c

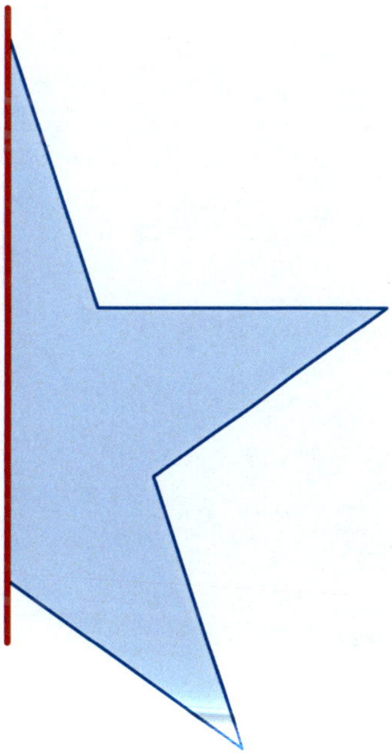

Practising my

maths language

Draw the line of symmetry in each of these pictures. Tell an adult how you knew where to draw the line of symmetry for each. The following sentence starters will help you.

This is the line of…

This side is reflected along…

You can see the reflection…

I know where it is because…

Use the words below to help you.

line of symmetry or line of reflection **reflection**

reflected **reflective symmetry** **mirror line**

a

b

c

d

e

From school to home

Find someone at home to talk to.

1. **Show them your school work. Talk about some of your answers to the questions on symmetry.**
2. **Explain what you have learnt about reflective symmetry.**
3. **Ask someone at home to help you complete the following:**

a What do you like about reflective symmetry?

____

____

____

b What was hard about reflective symmetry?

____

____

____

I understand | I understand a bit | I need more time to understand this

4 **How well do you understand this topic now? Circle one face for each statement.**

- I can draw reflections of shapes.
- I can draw a line of symmetry.
- I can identify lines of symmetry in everyday objects.

5 **Ask an adult at home to read and sign this.**

I have checked learning on this topic.

We have shared some understanding at home.

Signed ______________________________

Date __________

Teacher comment ______________________

____

Date __________

6 Data handling

Bar graphs

Padma looked at the colour of the cars that drove past her house one afternoon.

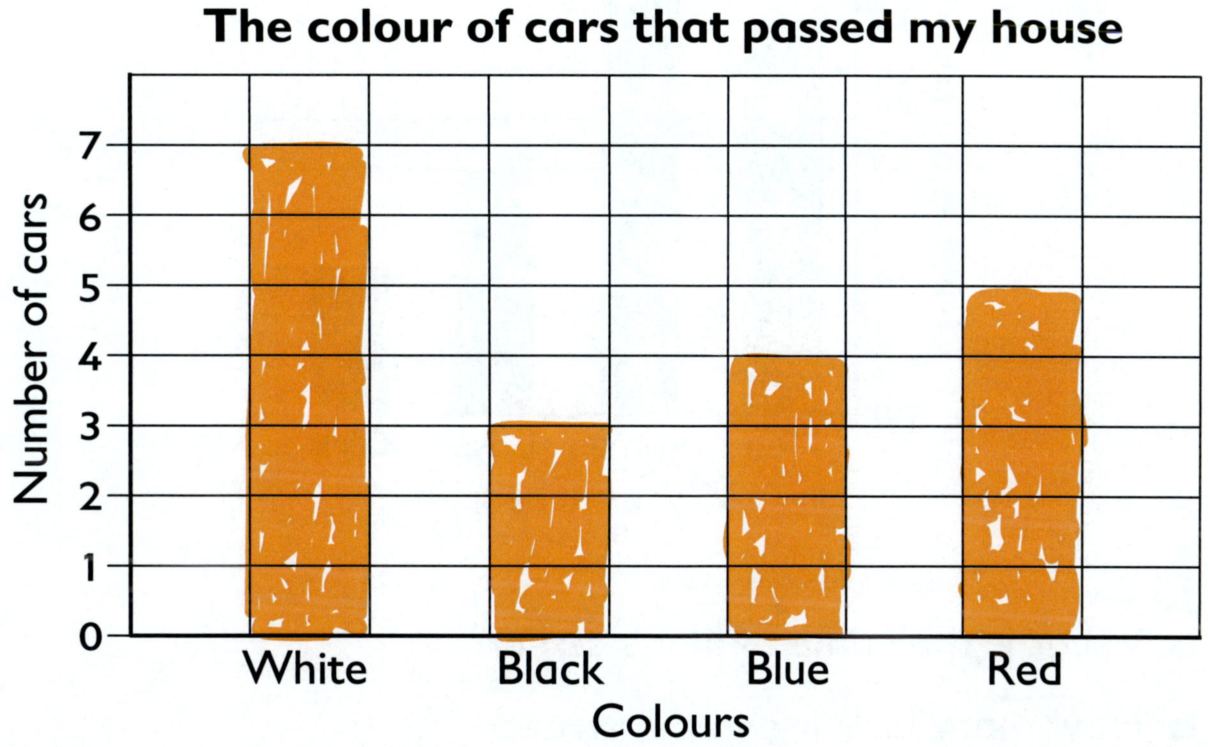

Talk to an adult, and step through what Padma did to create this bar graph. Tell them about the graph too, for example the title, vertical axis, horizontal axis, and axis labels.

Perhaps you could create a similar bar graph of cars that go past your house.

Check your maths!

The bar graph below shows the number of people there are in Han's street.

a What is the name of Han's street?

b How many girls live in the street?

c How many more women live in the street than men?

d How many children live in the street?

e How many people live in the street altogether?

Practising my

maths language

Look at the graph. Label these features:

title

axis label (there are two)

bar

vertical axis

horizontal axis

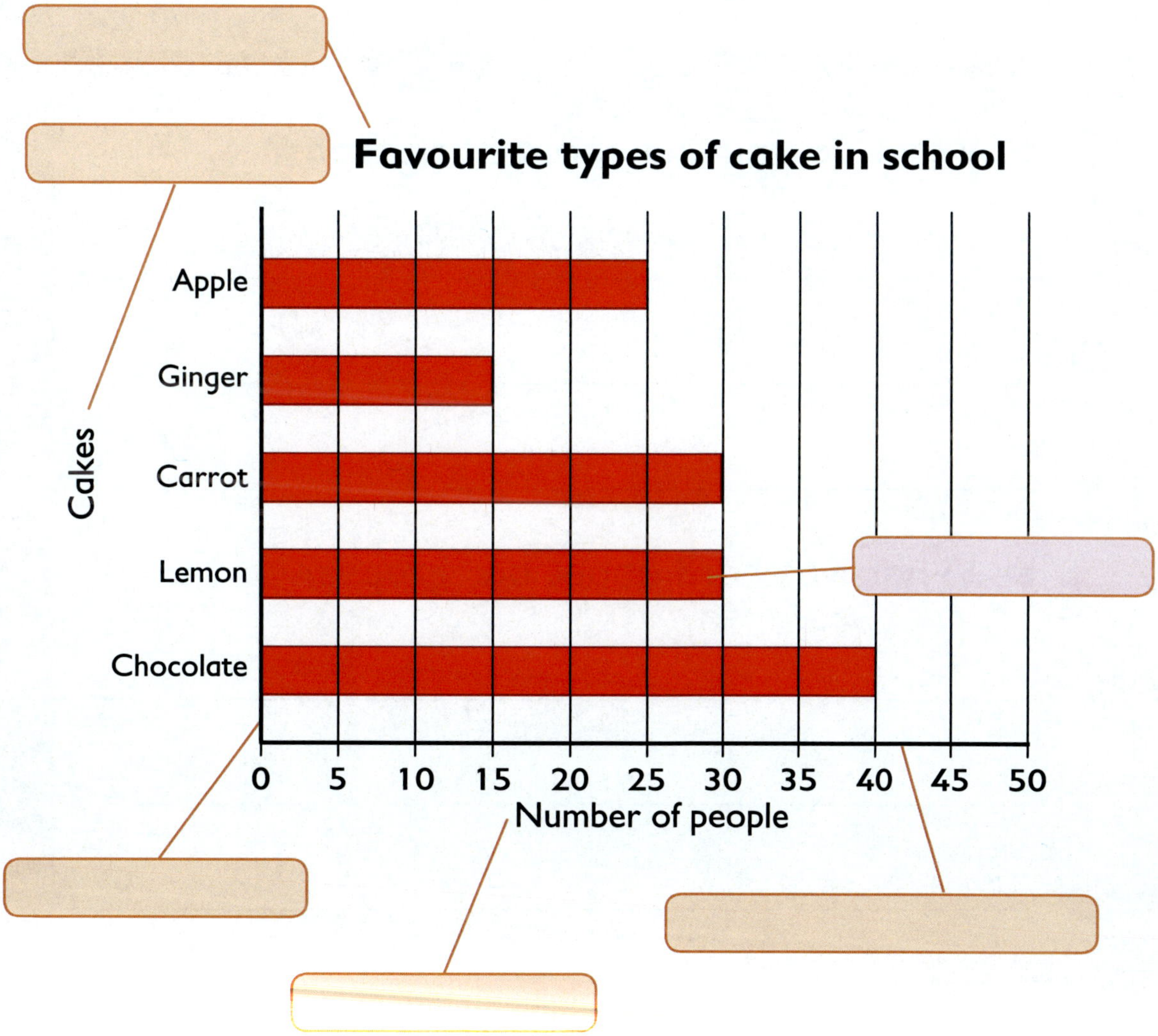

From school to home

Find someone at home to talk to.

1. **Show them your school work. Talk about some of your answers to the questions on data handling.**
2. **Explain what you have learnt about bar graphs.**
3. **Ask someone at home to help you complete the following:**

a What do you like about bar graphs?

____

____

____

b What was hard about bar graphs?

____

____

____

I understand | I understand a bit | I need more time to understand this

4 **How well do you understand this topic now? Circle one face for each statement.**

- I can identify the features of bar graphs.
- I can answer questions about data presented in bar graphs.

5 **Ask an adult at home to read and sign this.**

I have checked learning on this topic.

We have shared some understanding at home.

Signed ______________________________

Date __________

Teacher comment ______________________

____

Date __________

Sorting data using Venn diagrams and Carroll diagrams

Look at the ways the children sort the numbers.

Tya has sorted the numbers into even numbers and multiples of 3 using a Venn diagram.

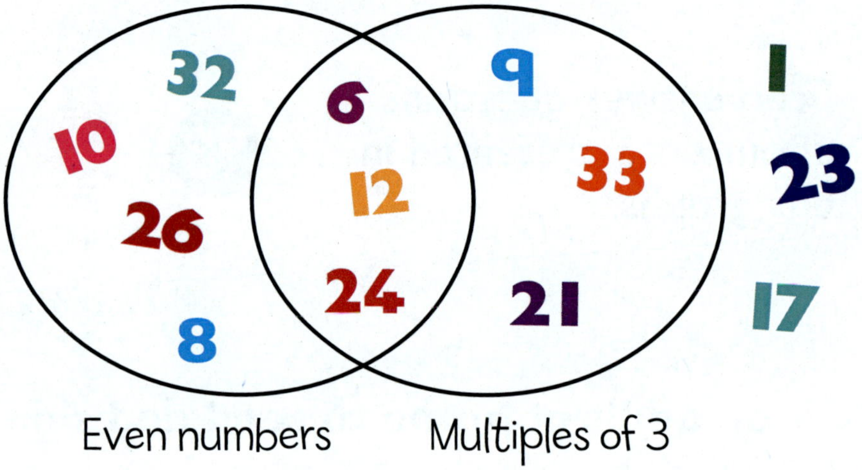

Han has sorted the numbers into even numbers and multiples of 3 using a Carroll diagram.

	Even numbers	Not even numbers
Multiples of 3	6 12 24	9 21 33
Not multiples of 3	8 10 26 32	1 17 23

Where would you put the numbers 18, 25 and 40 in each diagram?

Check your maths!

Draw the following shapes into the correct place on the Venn diagram.

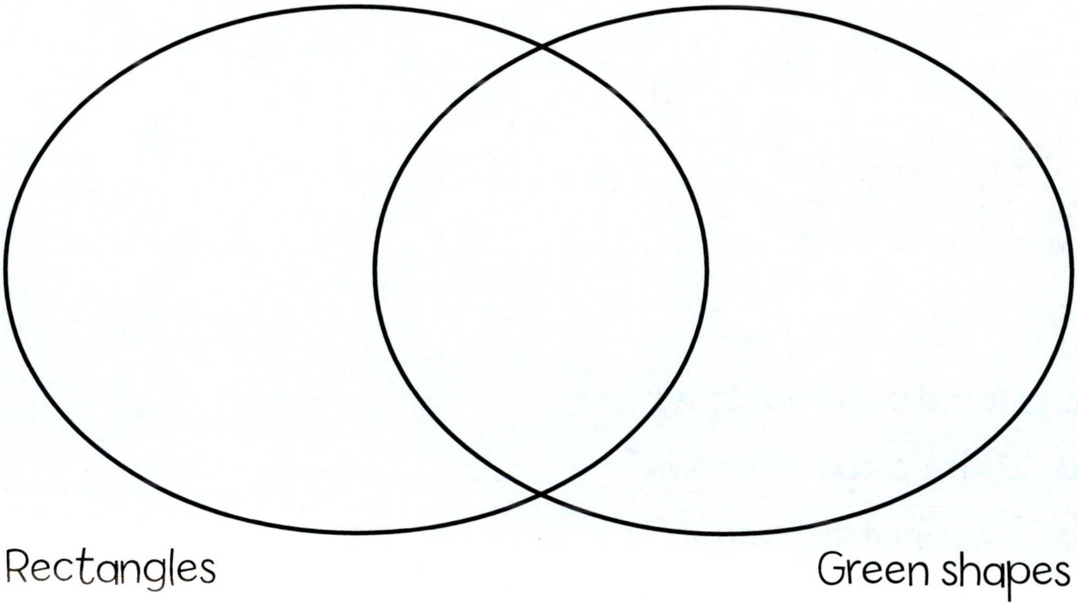

Practising my maths language

1 Write the correct labels in the spaces in the Carroll diagram. Remember to use 'not' in some of the labels.

	16 20 44 50	2 6 14
	17 23 41 55	1 3 7 13 15

2 Explain to an adult why:

a 20 belongs in its set

b 3 cannot belong in the same set as 23.

From school to home

Find someone at home to talk to.

1. **Show them your school work. Talk about some of your answers to the questions on data handling.**
2. **Explain what you have learnt about Venn diagrams and Carroll diagrams.**
3. **Ask someone at home to help you complete the following:**

a What do you like about Venn diagrams and Carroll diagrams?

____

____

____

b What was hard about Venn diagrams and Carroll diagrams?

____

____

____

I understand | I understand a bit | I need more time to understand this

4 **How well do you understand this topic now? Circle one face for each statement.**

- I can use Carroll diagrams to sort numbers or objects.
- I can use Venn diagrams to sort numbers or objects.
- I can explain choices using a range of words, including 'not'.

5 **Ask an adult at home to read and sign this.**

I have checked learning on this topic.

We have shared some understanding at home.

Signed ______________________________

Date ____________

Teacher comment ______________________

____

Date ____________

Multiplying teen numbers

We can use blocks to multiply teen numbers by 3 or 5.
We are going to find 3 × 14.

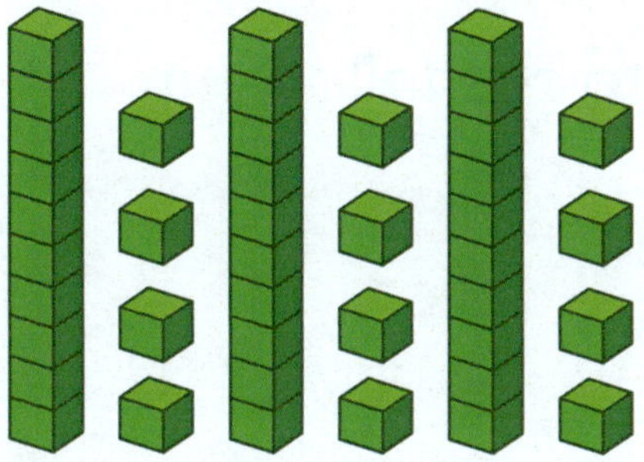

First, we make three sets of 14.

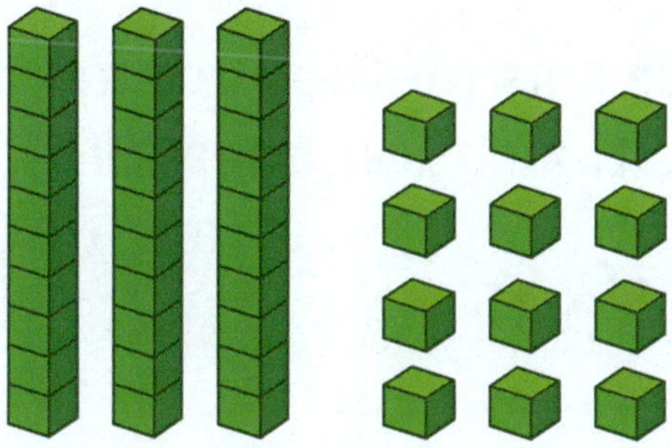

Now we can see that there are 3 tens and 12 ones.

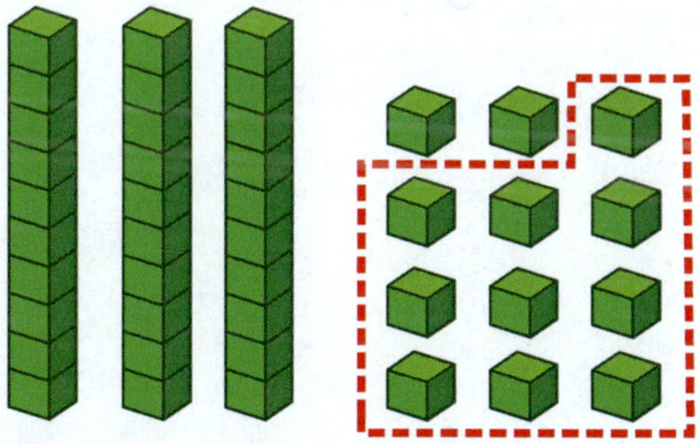

We can exchange 10 of the ones for 1 ten.

There are 42 altogether.
So, 3 × 14 = 42.

Would you have found the solution the same way? Explain to an adult the steps you would take.

Check your maths!

We can write 3 × 14 like this.

$$3 \times 14 = 14 + 14 + 14$$
$$= 10 + 10 + 10 + 4 + 4 + 4$$
$$= 30 + 12$$
$$= 42$$

Use this method to find the solution to the following multiplication questions.

a 5 × 16

b 3 × 12

c 3 × 18

d 5 × 19

Practising my

maths language

We are multiplying teen numbers by 3 or by 5. Talk to someone at home and tell them how you are using your knowledge of place value to multiply. Use these words.

multiplication facts **teen numbers** **sets**
product **counting blocks** **tens** **ones**

Here are some sentence starters to help you:

When I multiply teen numbers I think about the number using…

I need to know what happens when we multiply by ten because…

I always check the product by…

I use counting blocks by…

From school to home

Find someone at home to talk to.

1. **Show them your school work. Talk about some of your answers to the questions on multiplying.**
2. **Explain what you have learnt about multiplying teen numbers.**
3. **Ask someone at home to help you complete the following:**

a What do you like about multiplying teen numbers?

____

____

____

b What was hard about multiplying teen numbers?

____

____

____

I understand — I understand a bit — I need more time to understand this

4 **How well do you understand this topic now? Circle one face for each statement.**

- I can use counting blocks to represent teen numbers being multiplied by 3 or 5.
- I can multiply teen numbers by 3 or 5.

5 **Ask an adult at home to read and sign this.**

I have checked learning on this topic.

We have shared some understanding at home.

Signed ______________________________

Date ____________

Teacher comment ______________________________

____

Date ____________

8 Division

Division

Five children were asked to draw what they think about when they are asked to show how 2, 7 and 14 are related. Here is what they each came up with.

$2 + 2 + 2 + 2 + 2 + 2 + 2 = 14$

$2 \times 7 = 14$

$7 + 7 = 14$

$7 \times 2 = 14$

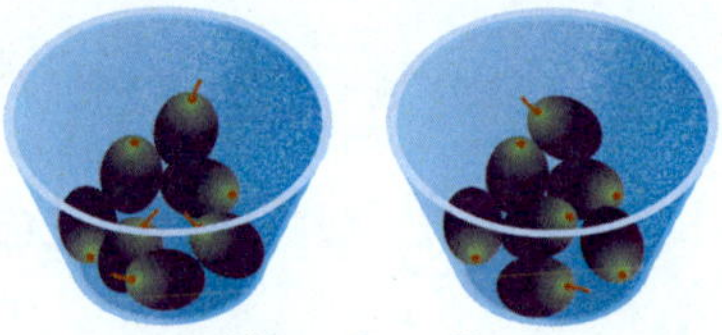

$14 \div 7 = 2$

$14 \div 7 = 2$

×	1	2	3	4	5	6	7	8
1	1	2	3	4	5	6	7	8
2	2	4	6	8	10	12	14	16
3	3	6	9	12	15	18	21	24

If $2 \times 7 = 14$, then $14 \div 7 = 2$ and $14 \div 2 = 7$ too!

Look at the different ways the children are thinking about multiplication and division. Are any of these ways your preferred way of thinking too? Or do you have another way to think about division? Explain your preferred way to an adult.

Check your maths!

Complete the number sentences using the multiplication facts to help you.

a

$6 \times 3 =$ 18

$3 \times 6 =$ 18

$18 \div 3 =$ ☐

$18 \div 6 =$ ☐

b

$4 \times 5 =$ ☐

$5 \times 4 =$ ☐

$20 \div 4 =$ ☐

$20 \div 5 =$ ☐

c

$2 \times 9 =$ ☐

$9 \times 2 =$ ☐

$18 \div 9 =$ ☐

$18 \div 2 =$ ☐

Practising my maths language

Sometimes you can use your multiplication facts to divide, and sometimes you can share evenly. Look at each picture showing equal grouping and write the division sentence in words and in numbers. The first one has been done for you.

a

We share 24 cookies equally on 6 plates. $24 \div 6 = 4$

b

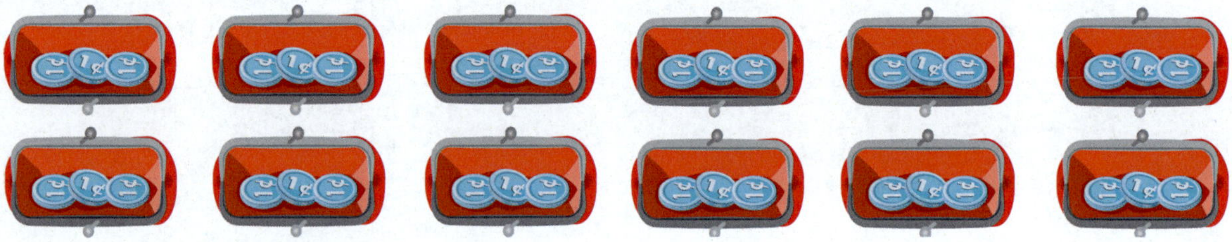

We share 36 cents.

c

d

From school to home

Find someone at home to talk to.

1. **Show them your school work. Talk about some of your answers to the questions on division.**
2. **Explain what you have learnt about division.**
3. **Ask someone at home to help you complete the following:**

a What do you like about division?

____

____

____

b What was hard about division?

____

____

____

I understand

I understand a bit

I need more time to understand this

4 How well do you understand this topic now? Circle one face for each statement.

- I can use multiplication facts to find division answers.
- I can use sharing to answer division questions.

5 Ask an adult at home to read and sign this.

I have checked learning on this topic.

We have shared some understanding at home.

Signed ______________________________

Date ____________

Teacher comment ______________________________

____

Date ____________

9 Length, mass and volume

Converting between lengths

The children are measuring their height in centimetres. They are then converting from centimetres to metres and centimetres.

The children are talking about how far they travel to school in metres. They are then converting from metres to kilometre and metres.

Check your maths!

Convert these lengths.

a 340 cm = ☐ m ☐ cm

b 5 m 37 cm = ☐ cm

c ☐ m = 5 km 600 m

d ☐ km ☐ m = 3 250 m

e 3 005 m = ☐ km ☐ m

f 6 m 20 cm = ☐ cm

Practising my

maths language

1 Match each abbreviation to the full term.

cm •	• metres
km •	• centimetres
m •	• kilometres

2 Write the best unit to find the measurement of the following. Use centimetres, metres or kilometres.

a The distance from your home to the nearest large city.

b The length of a pencil.

c The length of your hand.

d Your height.

e The distance from your home to school.

f The width of the main room in your house.

From school to home

Find someone at home to talk to.

1. **Show them your school work. Talk about some of your answers to the questions on length.**
2. **Explain what you have learnt about converting between lengths.**
3. **Ask someone at home to help you complete the following:**

a What do you like about converting between lengths?

____

____

____

b What was hard about converting between lengths?

____

____

____

I understand

I understand a bit

I need more time to understand this

4 **How well do you understand this topic now? Circle one face for each statement.**

- I can explain how many centimetres are in one metre.
- I can explain how many metres are in one kilometre.
- I can convert between kilometres and metres, and also between metres and centimetres.

5 **Ask an adult at home to read and sign this.**

I have checked learning on this topic.

We have shared some understanding at home.

Signed ______________________________

Date __________

Teacher comment ______________________

____

Date __________

10 Time

Reading the time

Look at the clocks in these pictures.
Tell an adult what time each event is happening.

a

b

c

d

Check your maths!

Match the clock with the correct time.

Three thirty-five

Half past seven

Twelve o'clock

Five minutes past ten

Four ten

Six twenty-five

Two twenty

Practising my

maths language

1 Circle the word in the sentences that means the sentence makes sense.

a It takes 5 minutes / seconds to boil an egg.

b Tya walks to school in 10 days / minutes.

c Toby can bounce a ball 10 times in 15 months / seconds.

d It takes less than one minute / month to say the alphabet.

2 Wear a watch or carry an analogue clock around the house with you today. Look at the time that you start something and the time that you finish, such as eating your dinner, brushing your teeth or reading a book.

From school to home

Find someone at home to talk to.

1. **Show them your school work. Talk about some of your answers to the questions on time.**
2. **Explain what you have learnt about reading the time.**
3. **Ask someone at home to help you complete the following:**

a What do you like about reading the time?

____

____

____

b What was hard about reading the time?

____

____

____

I understand | I understand a bit | I need more time to understand this

4 **How well do you understand this topic now? Circle one face for each statement.**

- I can read the time on analogue clocks to the nearest 5 minutes.
- I can choose the best unit of time to measure how long it takes to do something.

5 **Ask an adult at home to read and sign this.**

I have checked learning on this topic.

We have shared some understanding at home.

Signed ______________________

Date __________

Teacher comment ______________________

____

Date __________

11 Position and movement

Position, direction and movement

It is sports day at school and the children are running around the track.

Tya is running around the field clockwise.

Samir is running around the field anticlockwise.

Oh no! Tya and Samir have collided!

Explain to an adult at home why Tya and Samir have collided. Use the words **clockwise** and **anticlockwise** to help you.

Check your maths!

Where are the objects in this table?

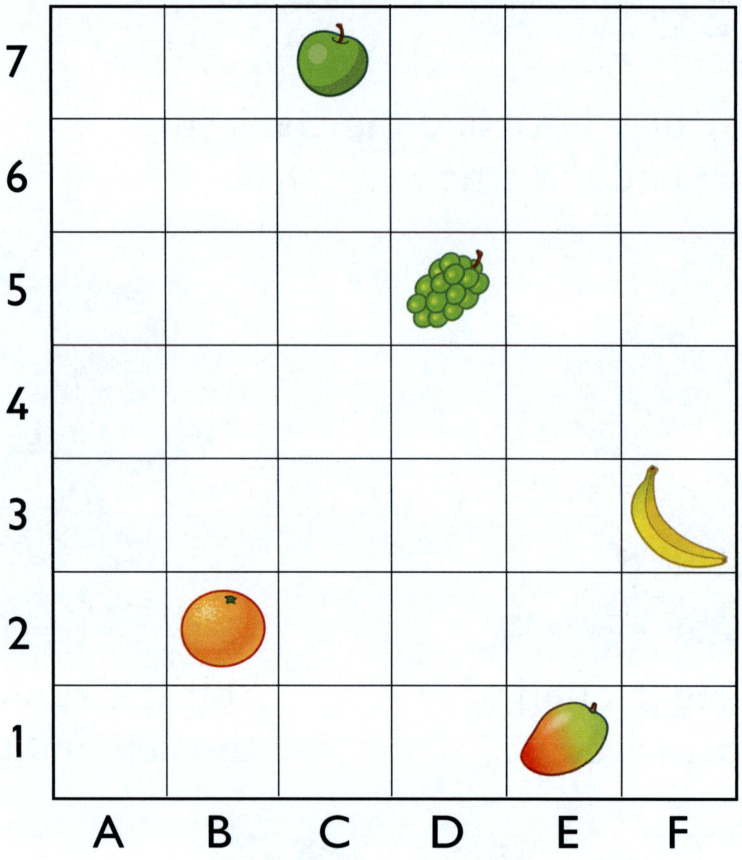

a The grapes are in ____.

b The orange is in ____.

c The apple is in ____.

d The banana is in ____.

e The mango is in ____.

Practising my maths language

We use different words to describe our position, direction or movement. Look at the pictures then complete the sentences using the words below.

over **under** **on** **outside** **inside** **clockwise** **anticlockwise**

a The bird is ______ the branch.

b Toby is ______ the house.

c Han is jumping ______ the water.

d The carousel is moving ______.

e The gift is ______ the box.

f Samir is ______ the table.

g Tya is turning the key ______.

From school to home

Find someone at home to talk to.

1. **Show them your school work. Talk about some of your answers to the questions on position, direction and movement.**
2. **Explain what you have learnt about position, direction and movement.**
3. **Ask someone at home to help you complete the following:**

a What do you like about position, direction and movement?

____

____

____

b What was hard about position, direction and movement?

____

____

____

I understand

I understand a bit

I need more time to understand this

4 **How well do you understand this topic now? Circle one face for each statement.**

- I can use words to describe position, such as on, inside, underneath.
- I can use words to describe direction, such as clockwise and anticlockwise.
- I can find and describe the position of objects on a grid.

5 **Ask an adult at home to read and sign this.**

I have checked learning on this topic.

We have shared some understanding at home.

Signed ______________________________

Date __________

Teacher comment ______________________

____

Date __________

Right angles

Jade has been using a set square to find right angles.

Look around you. How many right angles can you see?

Check your maths!

Write 90° next to the right angles.

Write 180° next to the straight lines.

Draw a cross next to the angles that are not right angles or straight lines.

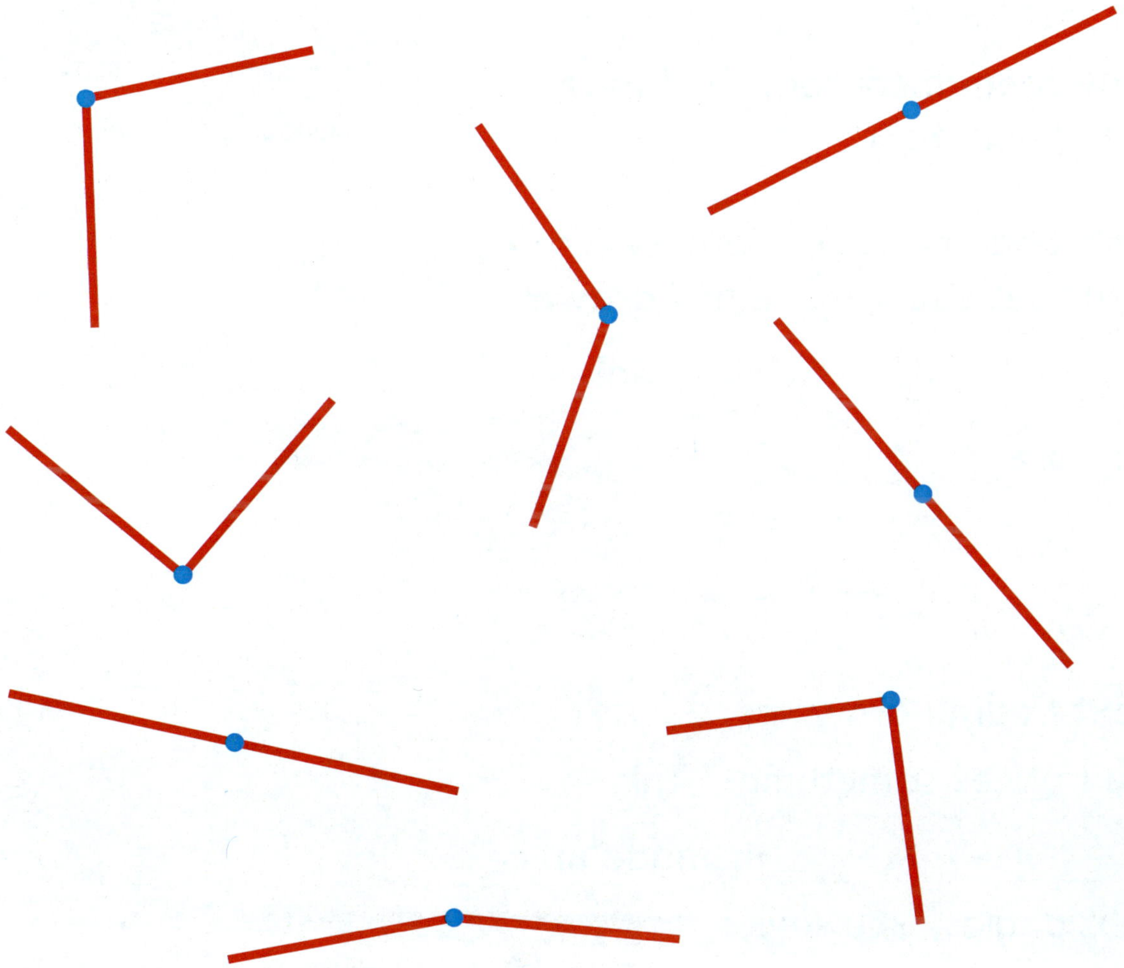

Practising my

maths language

Read the story below, then use these words to complete the sentences.

two   set square   straight line   right angle   90°

Padma draws some angles

Padma wants to draw some angles.

She has found a ______ (a mathematical tool she knew would help her).

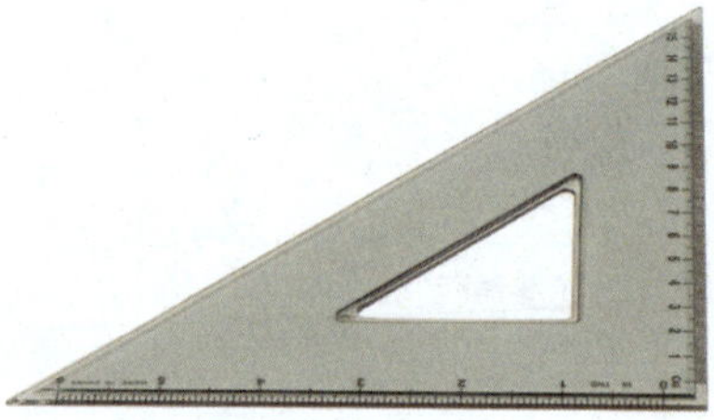

First, Padma draws along two sides of the set square. She has drawn a ______. This angle measures ______.

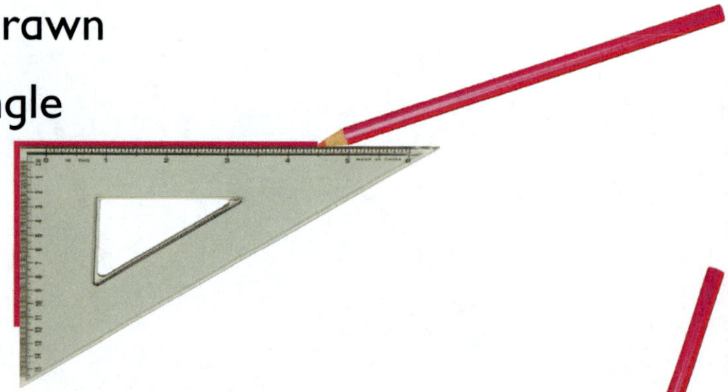

Next Padma draws a ______. She notices something. "This ______ is made up of ______ right angles," she says.

From school to home

Find someone at home to talk to.

1. **Show them your school work. Talk about some of your answers to the questions on angles.**
2. **Explain what you have learnt about right angles and straight lines.**
3. **Ask someone at home to help you complete the following:**

a What do you like about right angles and straight lines?

____

____

____

b What was hard about right angles and straight lines?

____

____

____

I understand

I understand a bit

I need more time to understand this

4 **How well do you understand this topic now? Circle one face for each statement.**

- I can identify right angles and know they measure 90°.
- I can identify straight lines and explain how two right angles make a straight line.

5 **Ask an adult at home to read and sign this.**

I have checked learning on this topic.

We have shared some understanding at home.

Signed ______________________________

Date ____________

Teacher comment ______________________

____

Date ____________